W9-AUU-326

Published by Creative Education
123 South Broad Street, Mankato, Minnesota 56001
Creative Education is an imprint of The Creative Company

Designed by Stephanie Blumenthal
Production Design by Melinda Belter

Photographs by: Frederick Atwood, Earth Images, FPG International,
KAC Productions, Peter Arnold, Inc., James H. Robinson, and
Tom Stack & Associates

Library of Congress Cataloging-in-Publication Data

Richardson, Adele, 1966–
Seashells / by Adele Richardson
p. cm. — (Let's Investigate)
Includes glossary.
Summary: Examines how shells are formed, the animals that produce them,
and what is involved in starting a collection.
ISBN 0-88682-996-8
1. Shells—Juvenile literature. 2. Mollusks—Juvenile literature.
[1. Shells. 2. Mollusks.] I. Title. II. Series: Let's Investigate (Mankato, Minn.)
QL405.2.R52 1999
594'.1477—dc21 98-43734

First edition

2 4 6 8 9 7 5 3 1

Creative Education

SHELL FACT

A shell will increase in size and strength as the animal it belongs to grows.

Conch

You can find shells as you walk along the seashore, down the banks of lakes and rivers, or even in a forest. Some are large, shiny, and colorful, while others are so small they can hardly be seen. But all shells are fascinating wonders of nature.

LIVING SHELLS

When a shell is found washed up on the seashore it is usually empty. But once it was the home of a sea animal. Many animals, such as turtles, have shells. Usually when people speak of shells, however, they mean those soft-bodied animals called **mollusks.**

SHELL
THICKNESS

In most cases, mollusks that live in deeper water will have thinner shells and hardly any color.

Scallops

SHELL

INSTRUMENT

The false trumpet is the largest sea snail. It lives in the waters around Australia and can grow up to two feet (61 cm) long.

Mollusks are a large group of animals containing as many as 100,000 living species and more than 50,000 fossil forms. This number will continue to grow as scientists are able to explore deeper into the ocean. Most mollusks known today can be divided into five main groups: **univalves, bivalves,** tooth-shells, octopuses and squid, and **chitons.**

Right, inside a chambered nautilus; far right, Florida worm shell

Most mollusks, but not all, have their shells on the outside of their bodies. The shell, the skeleton of the mollusk, is very strong, like a suit of armor that protects the soft animal. The shell is actually a part of the mollusk and is attached to the animal by muscles.

SHELL SNACK

Cuttlebone is placed inside pet bird cages for a snack and beak sharpener.

SHELL FACT

If a mollusk ever leaves its shell, it can never go back and reattach itself to its skeleton.

SHELL FACT

The sea snail is only one-tenth of an inch (.25 cm) long when hatched. It will grow to five or six inches (13 or 15 cm) long in six months.

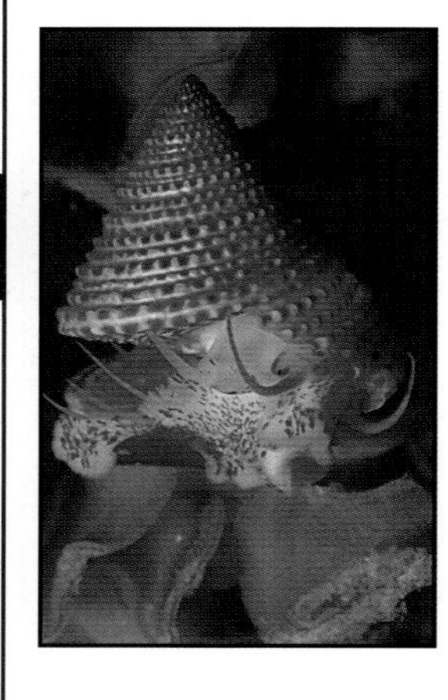

Some shells grow on the inside of animals. These shells are called **pens** when they are on the inside of squid. Inside cuttlefish they are called **cuttlebones.** They help to support the bodies of the animals, just like the human skeleton supports the body.

Above, purple ringed-top snail; right, Atlantic hairy Triton

SHELL FORMS

The shell, built by the mollusk itself, is made from a form of **lime-stone.** The mollusk has special **glands** that are able to take limestone from water and place it in tiny bits along the inside and edge of the shell. Most shells contain three layers: **prismatic, lamellar,** and **nacreous.**

SHELL

GIANT

The giant African snail has the largest shell of all land snails. It will grow to be about eight inches (20 cm) long.

Above, cuttlefish

SHELL TWIST

The snail's shell is a tube that twists and winds around itself as it grows.

SHELL COLOR

Tree snails usually have brighter and more colorful shells than water snails. This is due to the different foods the animals eat.

Right, close view of a scallop; far right, murex

The mollusk uses minerals in the foods it eats to form the shell and give it color. The minerals travel through the bloodstream to the animal's **mantle.** Inside the mantle are special glands that produce liquid materials to make the shell. There are also other glands with hardening materials so that the liquid will become solid and strong. Still more glands will produce the color.

SHELL FACT

There is no telling how long a shell will last because they do not decay and insects will not eat them.

Above, cowry; top right, helmet shell; bottom right, Triton's trumpet

UNIVALVES AND BIVALVES

Univalves are mollusks that have only one shell. The members of this group have a shell that may be cap-shaped or twisted around in a spiral. Out of the open part of the shell a "foot" is extended so the animal can move around. The foot is actually a part of the lower body of the animal.

Most snails have **right-handed shells.** This means the shells grow toward the right, or in a clockwise direction, when viewed from above. There are also a few types of **left-handed shells.**

SHELL
DOORS

Some shells have trap doors that a collector must save and glue back in place once the animal is removed.

SHELL
JEWEL

Mother-of-pearl, the shiny material found inside of nautilus shells, is often used to make jewelry.

Snail

SHELL

TINY

The smallest kinds of shells are those of marine snails. Some are as small as a grain of sand, even when fully grown.

One type of univalve, called the carrier shell snail, attaches tiny bits of stones or other objects to its shell as it grows. The hardening materials from its glands acts like cement and traps the objects in the shell. Once the shell is hard the objects cannot be removed unless the shell is broken.

Mole cowry

SHELL

WIDTH

The shell of a nautilus can grow to be as wide as 10 inches (25 cm) across.

Mollusks that are bivalves have two matching shells. Some animals in the bivalve group are clams, oysters, and scallops. The two shell halves open and close on hinges that look a lot like small teeth. When a bivalve is resting, its two halves are open. The animal has a wide, stretchy tissue that holds the two shells apart.

Giant clam in the Coral Sea

SHELL
COLOR

Once a mollusk has put color in a shell, it is permanent and will never fade.

SHELL
FOREST

Land snails are very common in damp forest areas, but they are fussy—they won't live in a pine forest!

Right, beached mussels; far right, this murex is being eaten by a magnificent tulip

There are also two strong muscles, called **adductor muscles,** attached to both shells. These muscles will snap the two halves closed if an enemy comes near. The shells will open again only if these muscles become tired. Most of the time an enemy does not wait for this to happen and leaves the animal alone.

SHELL

TRAP

The largest bivalve is the giant clam found in the Pacific and Indian oceans. It can grow to be four feet (122 cm) long and weigh 500 pounds (225 kg). Careless divers have been trapped by these giant creatures!

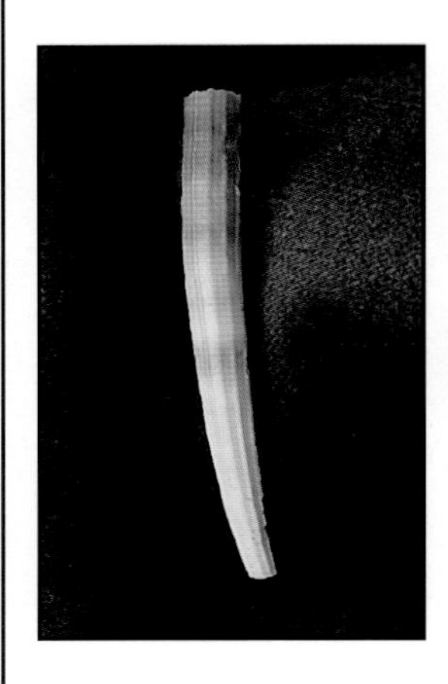

Above, tooth shell; right, common harp

SHELL VARIETIES

A "tooth-shell" mollusk has a shell that looks like a little elephant tusk. They are also called tusk shells. These shells are hollow tubes that are smaller at one end and slightly curved. Tooth-shells have openings at both ends. They can be found on the ocean floor all over the world with one end burrowed in the sand. These mollusks usually live in deeper waters.

The octopus and squid are in the group of mollusks that have shells on the inside of their bodies. The best known shell from this group is the chambered **nautilus.** This shell is divided into a series of chambers, or compartments, that become larger as the animal grows. When a new chamber is formed, the older, smaller chambers are sealed off. The outer one, where the animal is found, remains unblocked. The tentacles of the animal are extended from the head of the nautilus. This shell is found mostly in the western Pacific Ocean.

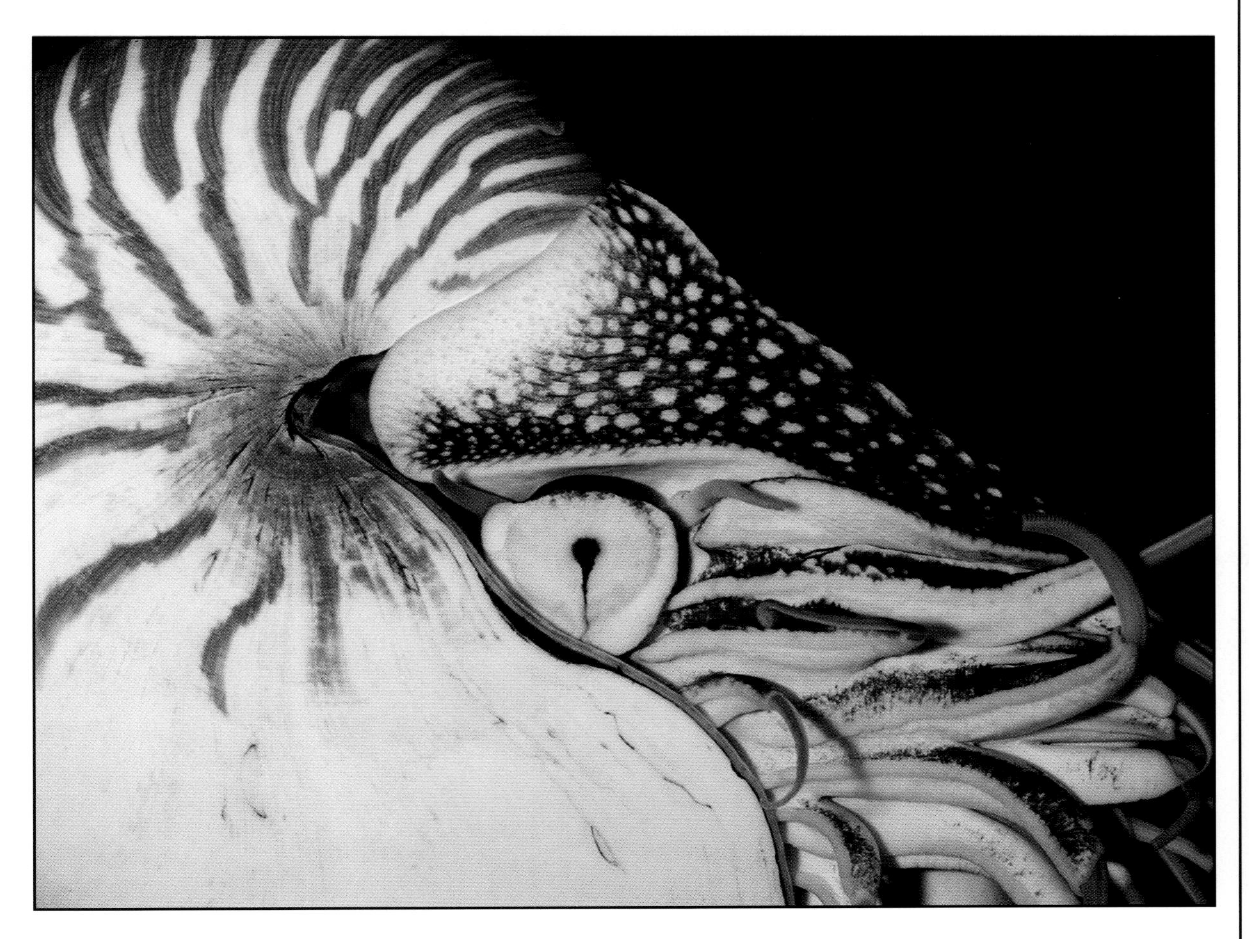

SHELL

SLIME

Land snails have glands that spit out slime in front of them so they can glide across the ground.

Right, a living chambered nautilus; below, marble cone

SHELL BABIES

A female oyster can produce almost half a billion eggs in one year, but only one baby in every four million will reach adulthood.

The chiton is a mollusk with eight single pieces, or plates, on its back for its shell. These plates are movable and are held together by a surrounding leathery belt. The belt acts like a string of hinges that allows the animal to bend and move around easily. The shells of chiton are also called coat-of-mail shells because they look like chain-mail from suits of armor.

Right, the eggs of shell creatures; far right, chiton

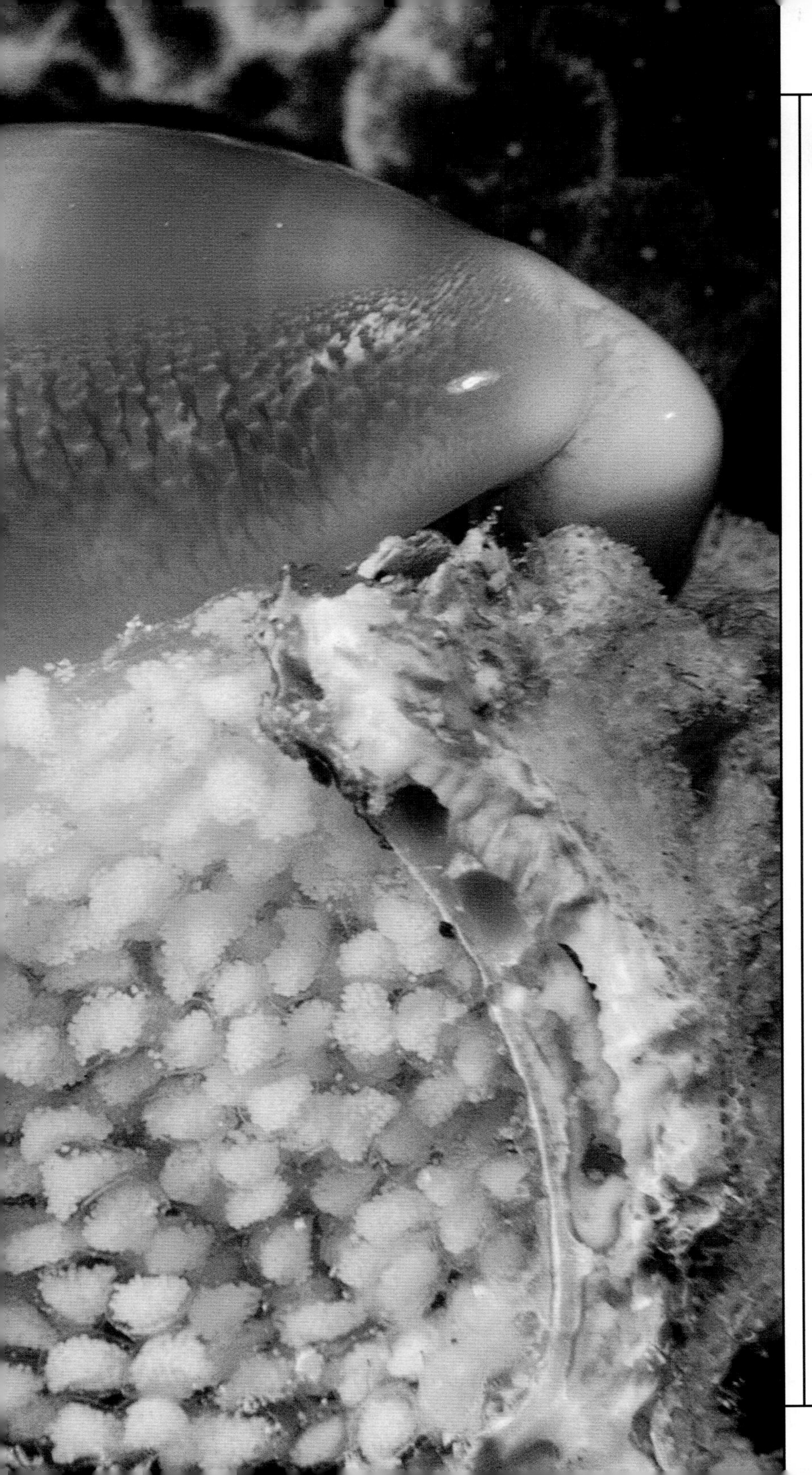

REPRODUCTION

Some species of mollusks, such as clams and snails, lay eggs to reproduce. The young will begin to grow their shells while still inside their egg. When the time comes for them to hatch, they emerge as perfect little copies of their adult parents.

SHELL CASH

Some early cultures used shells as a form of money.

SHELL DIP

Shell collectors have to remove the animals from tiny shells by soaking them in a container filled with 50 percent alcohol and 50 percent water for 24 hours.

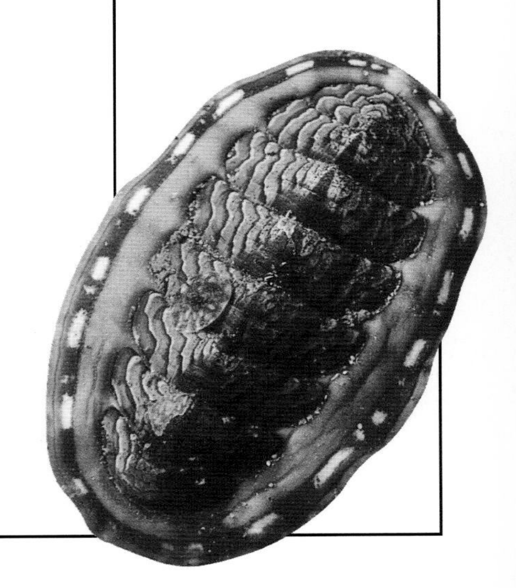

SHELL

PUMP

A large oyster can filter oxygen from more than 40 gallons (191 l) of water every day through its gills.

Above, octopus; right, this Triton's trumpet is feeding on a starfish

A few species of mollusks give birth to live young in much the same way that a dog or cat does. There are also some types of mollusks that have no males of their species. The females are able to produce the eggs, or the young, without the need of a male for fertilization. Scientists call this type of animal **parthenogenetic.** Some frogs, snakes, and lizards are also parthenogenetic.

Gulf coast seashells

MYTH

When you put a shell up to your ear you can hear the ocean.

TRUTH

What you hear is the sound of air movement being ***magnified*** *by the curves and twists inside the shell*

SHELL COLLECTING

Since there are so many different types of shells, collecting is a fun and interesting hobby. A good reason to consider a shell collection is that most shells require very little care and do not break easily.

SHELL FACT

Some mollusks have been found in water as deep as 2,200 feet (670 m)—that's more than one-third of a mile!

Above, lightning whelks; right, abalone

A hobbyist can find new additions for his or her collection just by walking along a beach. Other places to look for shells are in tide pools, between rocks, or maybe even on a shipwreck! Some collectors dig in the sand along the water's edge at low tide to find their prizes.

To build up your collection, gather several shells of the same type when you find them. This way you will have extra shells to trade with other collectors, and your own collection will grow faster.

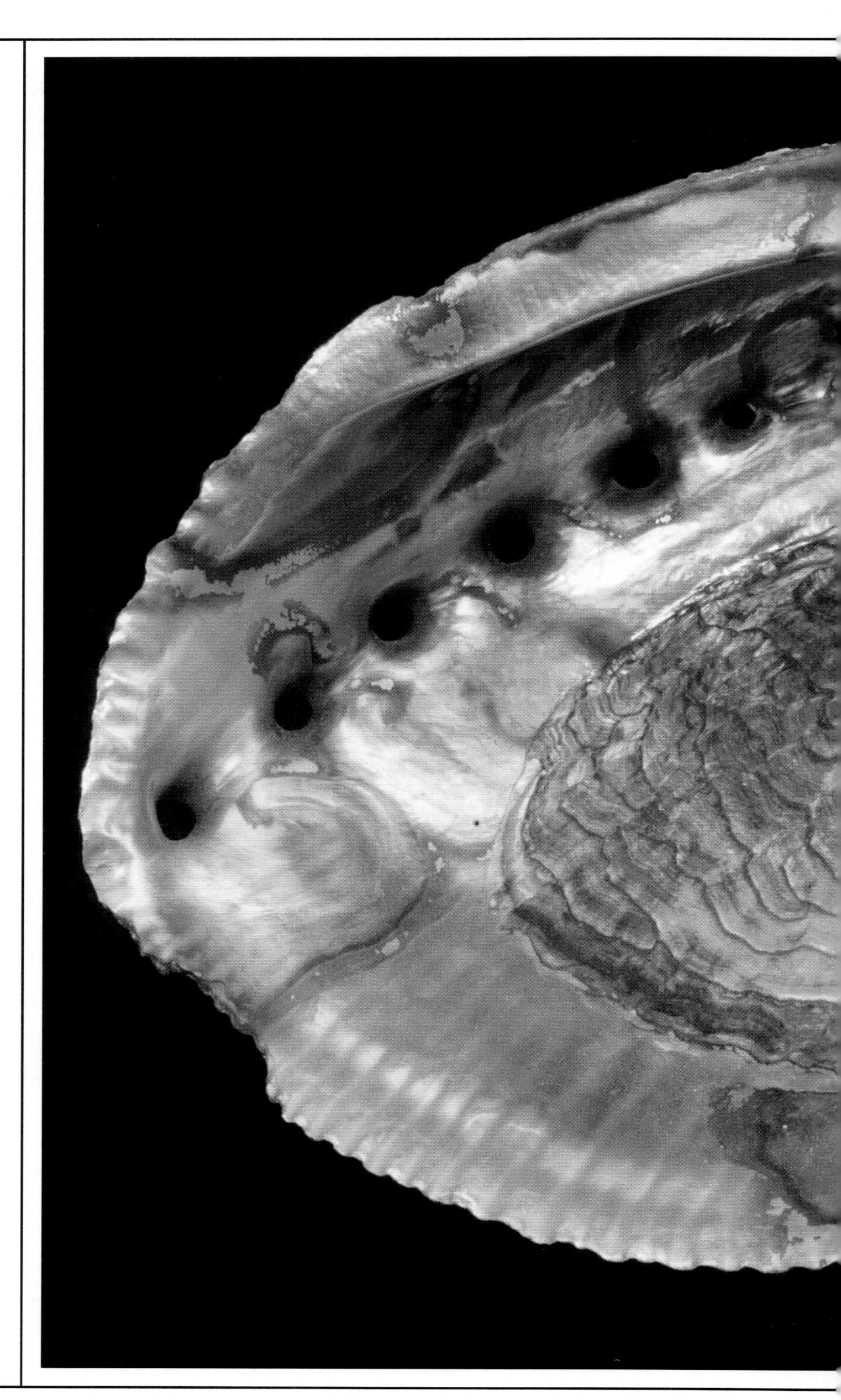

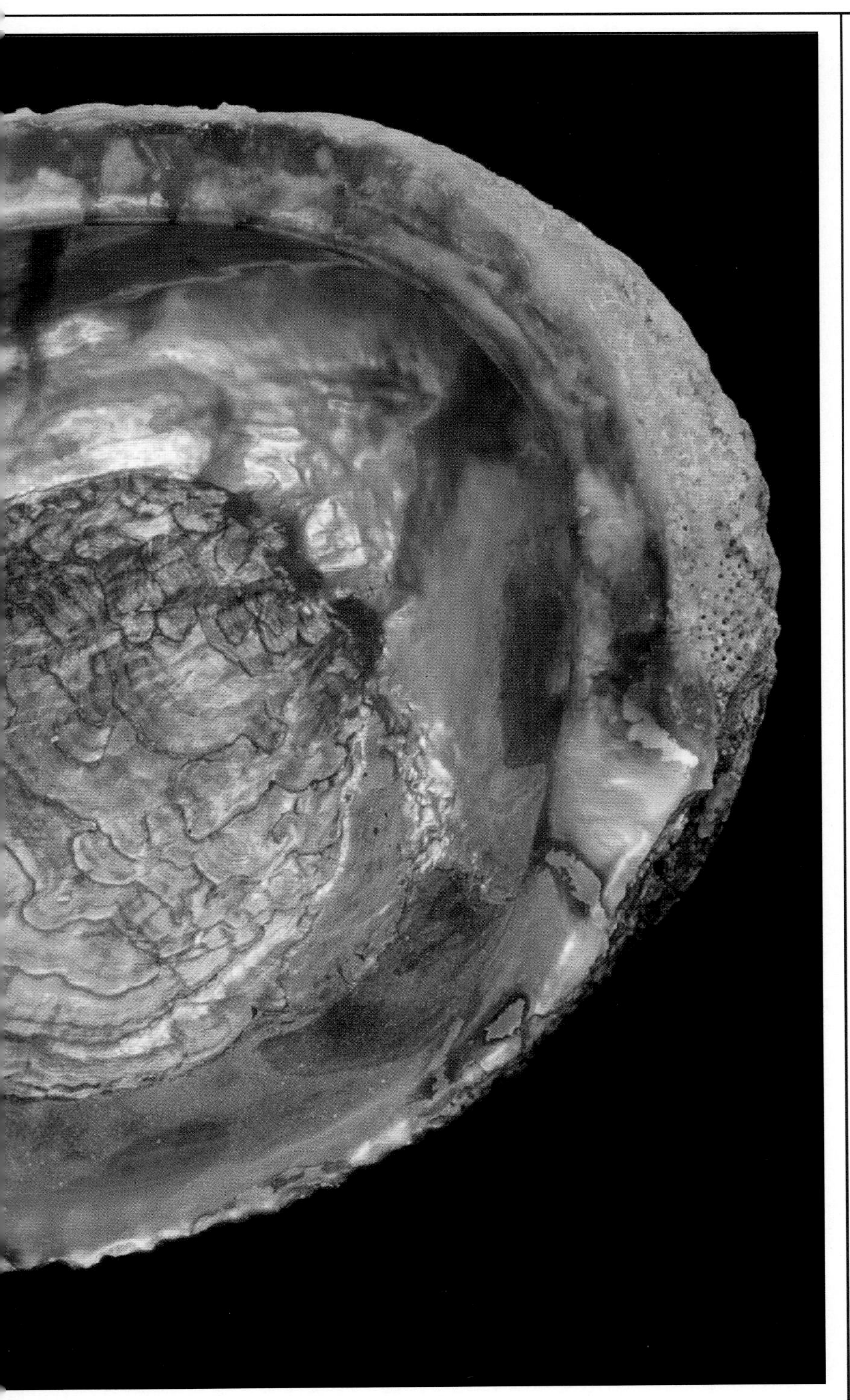

Don't forget about fresh-water snails. These are found in ponds, streams, and ditches. Even if you don't live near an ocean, you can still hunt for shells. Land snails also produce shells that can be added to a collection. They are found on the ground, in piles of old shrubbery, or even in rotted logs or tree stumps.

SHELL SIZE

The horse conch, found off the Florida coast, is the largest shell native to the United States. It can grow to be up to one and three-quarters feet (53 cm) long.

SHELL
LENGTH

The largest chiton shells are found off the California coast, growing up to one foot (30 cm) long.

SHELL
MONSTER

Giant squids have been reported in the Atlantic Ocean at lengths of more than 50 feet (15 m); they can weigh over 4,400 pounds (2,000 kg).

Right, tonna perdix; far right, a diver in the Philippines

SHELL CARE

The messy part of shell collecting comes when you find a shell with the animal still inside. If this happens you have to remove the animal or it will decay and start to smell bad.

To remove the animal, gently drop the shell into a pot of boiling water for a minute or two. This will, unfortunately, kill the animal, but it will keep a bad odor out of the house.

SHELL

GEM

When a grain of sand gets stuck in the mantle of some oysters and clams, the animal produces a pearl, which is highly valued by jewelry makers.

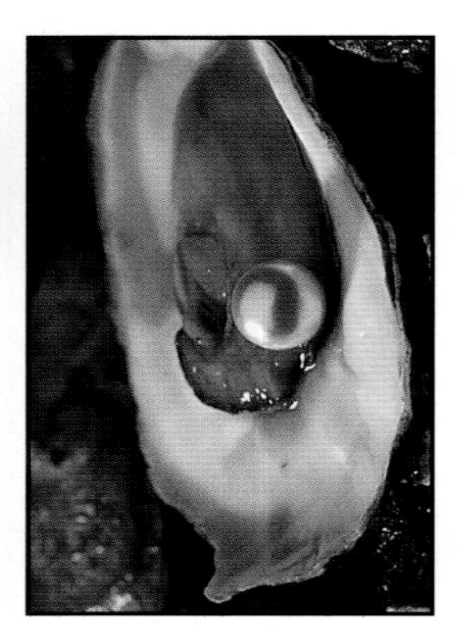

You must use tweezers to remove the dead animal from a univalve shell. To dry off the shell, let it sit in a shaded area for a day. If you towel dry it, you may damage the shell. Once the shell is completely dry, gently pack the inside with cotton.

The larger bivalve mollusk should be dropped into boiling water until the two halves open.

Above, oyster with its pearl; right, northern moon shells

Carefully take the shell out of the pot and remove the animal with a spoon or dull knife. Once the body is scraped out, but before the hinge dries and hardens, tie the two halves together with a string and set them in a shaded area. You must wait until the shells are completely dry before removing the string. If done correctly, the shells will stay closed. A thin covering on the outside of the shells may peel off. To keep this from happening, polish your collection with a light coating of vegetable oil.

SHELL FACT

None of the members of the bivalve group of shell animals can live out of water.

SHELL
SIZE

Most mollusks are very small and weigh less than one-fourth of an ounce (7 g).

Top right, periwinkles on a coral; bottom right, lion's paw scallop; far right, rooster conch

People don't need to have a collection to appreciate shells. We can learn about them and see many colorful pictures of all the world's shells in books and magazines. And the next time you're walking along the beach, or even in the woods, take a look around and see just how many different shells there are around you.

Glossary

Adductor muscles are muscles attached to both halves of bivalve shells. They are used to open and close the shell halves.

Bivalves are mollusks with two matching shells.

Chitons (KITE-ns) are mollusks with eight hard, strong protective plates on the back.

The shells providing body support inside cuttlefish are called the animals' **cuttlebones.**

Glands are organs that perform certain functions inside the body of a human or an animal. Bodies contain many different kinds of glands.

The **lamellar** is the middle layer of a shell.

Left-handed shells grow in a left, or counter-clockwise, direction when viewed from above.

Limestone is a type of calcium-rich rock that is formed by the dead remains of once-living animals.

When something is **magnified** it is made to look or sound large.

The fleshy tissue inside a shell is called the **mantle.** It has special glands for forming and coloring shells.

Mollusks are soft-bodied animals that grow shells for body support and protection.

The inside layer of a shell is the **nacreous.**

A **nautilus** is a mollusk with a chambered shell on the inside of its body.

Animals that reproduce without the need of a male for fertilization are called **parthenogenetic.**

Pens are shells on the inside of squid.

The outer layer of a shell is called the **prismatic.**

Right-handed shells grow in a right, or clockwise, direction when viewed from above.

Univalves are mollusks with one shell.